The Cerebro Standstill

A Book On Stroke And Stroke Management

By

Paolo Jose de Luna

Paolo Jose De Luna

Table of Contents

INTRODUCTION..................................3
Chapter 1 - What is Stroke?................6
   Onset and Progression......................9
   Epidemiology...................................10
   Risk Factors of Stroke.....................13
Chapter 2 - Classifications Of Stroke..16
   Ischemic Stroke...............................18
   Hemorrhagic Stroke........................22
   Pediatric Stroke..............................26
Chapter 3 - Signs And Symptoms Of Stroke..28
Chapter 4 - Diagnosing Stroke............39
Chapter 5 - Management of Patients Stroke..46
   Medical Management......................47
   Surgical Management......................53
CONCLUSION....................................55

INTRODUCTION

When the Brain Halts...

The brain is just one of the two most vital organs in the body, sharing the crown with the heart. The brain is responsible for a pyramid of functions of the body like memory, learning capacity, hearing, sight, touch, breathing, heart rate, temperature, hunger, and a whole lot more. Because of its functions, it's essential that the brain is kept healthy and alive. And just like the heart, the brain never ever stops working – ever. Even when you sleep, the brain is still "awake" and still up working. In fact, your dreams are being projected by the brain and your sleep-wake cycle is governed by the brain itself, so it never actually sleeps.

Being one of the most common neurologic health problems in the world, cerebrovascular accident or "stroke" for shorthand, has gained an alarming rise throughout the years. What we have to understand is that stroke isn't just a single disease, but rather, a group of conditions that lead to the situation wherein a cerebrovascular accident is likely to happen. Stroke is when the brain experiences a halt in its functions due to poor blood flow to the brain, resulting in the death of the brain cells.

The moment the brain stops working, even for just a second, the whole body feels a significant blow. Stroke can either happen because of the lack of blood flow to the brain due to vasoconstriction of the cerebral blood vessels or a clot that has clogged the flow of blood to the brain or because of a hemorrhage from a ruptured blood vessel or aneurysm, resulting in one or several parts of the brain to malfunction.

In the treatment of stroke, it's important that we know the details about it, especially on how to prevent it. There are some groups that are more prone to develop stroke in their lifetime, as old age and high blood pressure are two of the biggest risk factors for people to develop stroke. And stroke is no joke as without immediate intervention, it can result to lifelong disability and sometimes, maybe even death. And when it comes to the management of stroke, knowledge about it is important, not just for the person

The Cerebro Standstill

who has experienced stroke, but also for the family since the recovery period relies largely on the role of the family.

In this eBook, we'll be talking everything about stroke including what it really is, the different types of stroke, the signs and symptoms of cerebrovascular accident, the diagnostic tests used to diagnose stroke, the risk factors and causes of stroke, and the different treatment and management regimens involved in helping patients recover from stroke or cerebrovascular accident.

Are you ready? Get your pen and paper ready to take down some notes, here is everything you need to know about stroke and how to deal with it.

Chapter 1 - What is Stroke?

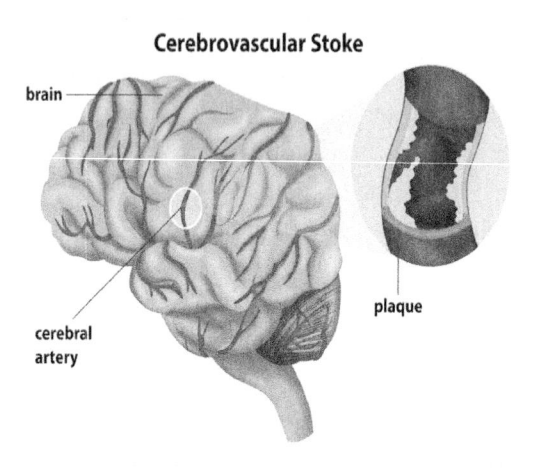

Cerebrovascular Stoke

Stroke, cerebro vascular accident (CVA), or brain attack occurs when there is a decreased blood flow to the brain which can result in brain cell death. This can happen because of two types of stroke – ischemic stroke which causes lack of blood flow to the brain due to ischemic changes to the cerebral blood vessels or due to hemorrhage because of a ruptured cerebral blood vessel, either way, these result in a part of the brain not functioning properly and the longer the

The Cerebro Standstill

brain doesn't receive enough blood flow, the more brain cells die with each passing seconds.

The thing that you have to understand first is that the brain is an ever active organ that needs a constant supply of oxygen and nutrition, in this case, through the blood. In just a few seconds of a decrease in cerebral blood flow, the brain immediately experiences the lack of oxygen and nutrients through the feeling of fainting, weakness, fatigue, or confusion. If those seconds turn into minutes, the signs and symptoms grow worse and the effect of the brain tissue becomes far graver since it risks the occurrence of brain cell death. After several minutes, the brain cells die and deliver a permanent effect of the brain which can exhibit things like paralysis, memory loss, behavioral changes, and maybe even death.

What's alarming in stroke is that the changes that occur depend upon the area affected by the decrease in blood flow, may it be the cerebrum, the cerebellum, or the brain stem. In any case, no matter what part of the brain is affected, stroke can still have debilitating effects to the body which can either be temporary or permanent.

The marking signs of an ischemic stroke are the inability to move one or more limbs, difficulty in feeling on one side of the body, difficulty in understanding or speaking, the feeling of fainting, or the loss of vision on one side. These signs and symptoms often occur after the stroke has occurred. If these signs and symptoms last for less than an hour, this type of stroke can be categorized as *transient ischemic attack* or TIA for short.

On the other hand, the marking sign of a hemorrhagic stroke is a severe headache and it usually leads to the deterioration in one's consciousness; if emergency treatment is not done, hemorrhagic stroke can lead to permanent neurologic changes, respiratory arrest, or even death.

Stroke can have a number of conditions associated with it because of the disability that it causes. Among them, health problems like pneumonia, decubitus ulcers, muscle atrophy, spinal problems, osteoarthritis, and urinary and fecal incontinence are the most common ones.

The Cerebro Standstill

Onset and Progression

The onset and progression of a stroke depends on its type. Ischemic stroke has a sudden onset without prior warning, usually beginning as an episode of mild paralysis in one or more limbs, difficulty in sensation, or difficulty in understanding or speaking. The progression of ischemic stroke then proceeds as slow, exhibiting neurologic deficits that depend upon the area of the brain that is affected. On the other hand, the onset of a hemorrhagic stroke is sudden, exhibited by an intensely painful headache that then descends the patient in a decreased or even loss of consciousness. From then on, the consciousness of the patient with hemorrhagic stroke will be difficult to stimulate and the progression of the condition to respiratory arrest is highly likely and always considered to be an emergency situation.

Epidemiology

Stroke is considered to be one of the most frequent causes of death in the world. In fact, just in 2011, it accounted for more than 6 million deaths. About 17 million people experienced a stroke in 2010 and more than 33 million people have previously had a stroke but were able to survive after treatment and rehabilitation. Throughout the years, the incidence of stroke has been decreasing by almost 10% in first world countries, while it has been increasing in third world countries. Overall, stroke has been experienced mostly by those exceeding the age of 65 years old.

The Cerebro Standstill

Stroke lies between heart disease and cancer when it comes to incidence and mortality rate. In the US, stroke is the leading cause of disability and it is the third cause of death. From 30 years of age, the incidence of stroke rises and the causes vary with age. Old age is one of the most important risk factors in determining the incidence of stroke and it has been found that 95% of strokes occur in those 45 years old and above, while 60% of strokes occur in those 65 years old and above. However, you should remember that stroke can occur at any age – even during childhood.

There's evidence of genetic tendency among those family members with stroke and sharing lifestyle habits that can contribute to the development of stroke.

An increased level of Von Willebrand factor is found to be more common to people who have had an ischemic stroke for the first time. This study then shows that even a person's blood type is a contributing factor to the incidence of stroke. Another good thing to note is that once you've had a stroke before, it increases your chances to develop strokes in the future as well.

When it comes to morbidity and mortality rates, men are 25% more prone to suffer from strokes compared to women, but 60% of deaths that occur because of stroke occur more in women. But this is only because women tend to live longer, thus increasing their risk in death when they experience in stroke. Other factors that make strokes worse for women include pregnancy, childbirth, menopause, and the treatment with HRT.

Risk Factors of Stroke

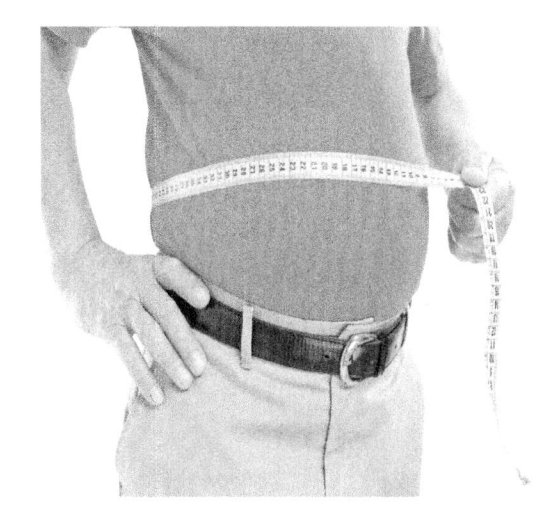

There are conditions that contribute to the development and incidence of stroke. These may include the following:

- Advanced age (30 years and older; more prevalent in the elderly)

- High blood pressure or hypertension
- Diabetes mellitus
- Smoking
- Heart diseases (e.g. ISD or ischemic heart disease, myocardial infarction)
- Obesity
- Previous stroke attacks
- Blood vessel dysfunctions or diseases
- Thrombocytosis or increased platelet count
- Clotting problems
- Familial history of stroke
- Hemorrhagic diseases (e.g. Hemophilia, Dengue Hemorrhagic Fever, etc.)
- Unrestricted inclusion of fat in the diet

- Hyperlipidemia or increased cholesterol levels
- Use of illegal drugs (e.g. cocaine, heroin, meth, etc.)

Chapter 2 - Classifications Of Stroke

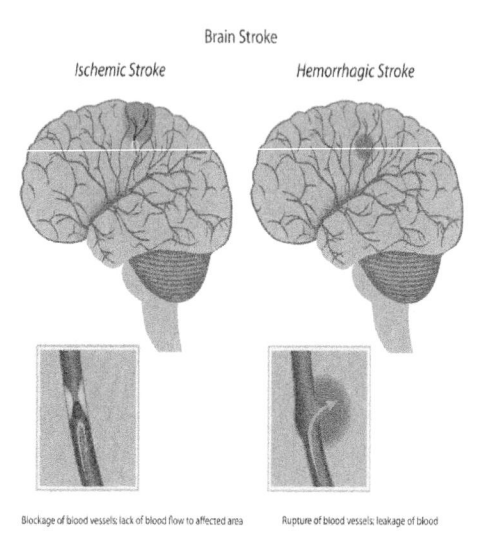

Brain Stroke

Ischemic Stroke | Hemorrhagic Stroke

Blockage of blood vessels; lack of blood flow to affected area | Rupture of blood vessels; leakage of blood

Stroke can be divided into two separate categories depending on their causes – ischemic stroke and hemorrhagic stroke. Ischemic stroke is caused by an interruption of the blood flow towards the brain, resulting in cell death of the brain cells if not treated immediately. On the other hand, a hemorrhagic stroke is caused by a bleed, often due to a ruptured blood vessel or aneurysm that then leads to blood leakage into the cranium and increases the pressure inside the skull.

About 80% of strokes are ischemic in nature, while the remaining 20% are hemorrhagic in nature. Those who have ischemic strokes in the past were also found out to have an increased risk to develop hemorrhagic strokes in the future, probably correlated to the weakening of the blood vessel walls and an increased risk to develop aneurysms and ruptured blood vessel structures.

Ischemic Stroke

Ischemic Stroke

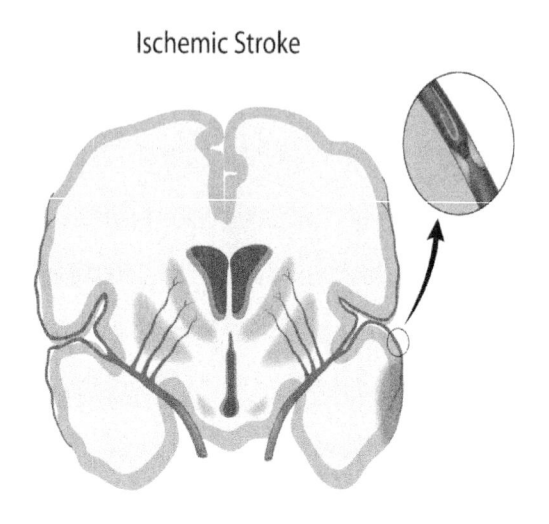

Ischemic stroke occurs when there is an interruption to the brain's blood flow. This can lead to a loss of function in one or more areas of the brain. As the interruption of the cerebral blood flow is prolonged, the damage to the brain cells worsen and if not treated immediately, it can leave the neurologic deficits that are

The Cerebro Standstill

supposed to be temporary only to risk them to become permanent.

Ischemic stroke can be due to four primary causes: a thrombosis or the obstruction of a blood vessel by a clot formation, an embolism or an obstruction of a blood vessel due to a clot that has traveled from another part of the body and then went to obstruct the cerebral blood flow, systemic hypoperfusion or a decrease in the overall blood circulation of the blood throughout the body like in the cases of shock, or venous thrombosis which is the formation of a blood clot in the dural venous sinuses which drains the blood from the brain. In any of these cases, interruption or a decrease in cerebral blood flow is the main culprit as to why patients with ischemic stroke deteriorate.

For strokes without a specific cause, the term "cryptogenic" is used to describe them. This type of stroke makes up for about 30-40% of all ischemic strokes in the world.

Various classification systems are used when it comes for ischemic stroke. One classification, the Oxford Community Stroke Project (OCSP) utilizes the initial symptoms of the stroke and depending on these symptoms; the ischemic stroke can be classified as a partial anterior circulation infarct or PACI, total anterior circulation infarct or TACI, lacunar infarct or LACI, or posterior circulation infarct or POCI. With these four classifications, the

The Cerebro Standstill

area and the extent of the stroke are determined, along with the potential cause and the prognosis for the patient.

Another classification for ischemic stroke is the TOAST (Trial of Org 10172 in Acute Stroke Treatment) classification. Unlike the OCSP classification, the TOAST utilizes the clinical symptoms on top of further investigations of the stroke. Using TOAST, the causes of the ischemic stroke is determined and may be due to a thrombosis or embolism from atherosclerotic changes of a large artery, an embolism that originated from the heart, a complete obstruction of a small blood vessel, other causes, and an undetermined cause.

Hemorrhagic Stroke

Hemorrhagic Stroke

Aneurysm

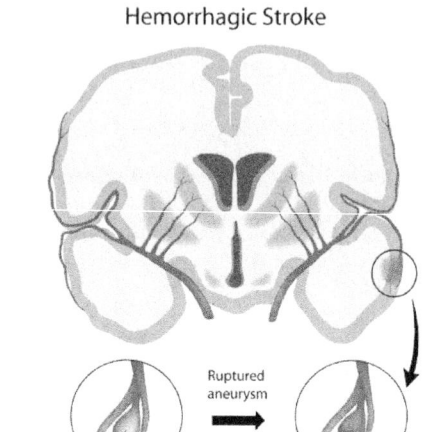

Ruptured aneurysm

The second type of stroke is much more serious compared to ischemic stroke. Hemorrhagic stroke is characterized by intracranial hemorrhage or the bleeding within the skull structure and cerebral hemorrhage or the bleeding within the brain tissue.

The Cerebro Standstill

Intracranial hemorrhage can be an epidural hematoma or bleeding that occurs between the dura mater and the skull, a subdural hematoma or bleeding in the subdural space, and subarachnoid hemorrhage or bleeding that occurs between the pia mater and the arachnoid mater.

Cerebral hemorrhage is the bleeding within the brain tissue and it can either be due to an intra-parenchymal hemorrhage or intra-ventricular hemorrhage (bleeding in the brain's ventricles).

Hemorrhagic strokes often begin as an episode of a severe headache that may be accompanied by vomiting. Oftentimes, the level of consciousness may deteriorate and some patients may even go into a coma.

In both cases of intracranial hemorrhage and cerebral bleeding, they are considered as emergency situations as they increase the intracranial pressure in the cranial structure and crush the delicate brain tissue.

Ischemic strokes and hemorrhagic strokes vary because of their causes and how they occur. With the slow and steady progression of ischemic stroke, it's still manageable given enough time. However, hemorrhagic strokes are always considered as emergency cases because of its rapid onset and progression that can even lead to respiratory arrest or death.

The Cerebro Standstill

There are certain risk factors that increase the incidence of hemorrhagic strokes. These risk factors may include the following:

- Hypertension or elevated blood pressure
- Diabetes mellitus
- Excessive alcohol consumption
- Use of illegal drugs (e.g. methamphetamine, cocaine, heroin, marijuana)
- Cigarette smoking
- Menopause
- Previous episode of ischemic stroke
- Aneurysm formations
- Small blood vessel diseases
- Clotting problems
- Severe cases of migraine
- Head injuries

Pediatric Stroke

Unlike common belief, stroke can also happen in the young. When stroke occurs in children, it is termed as "pediatric stroke" and is similar to ischemic or hemorrhagic stroke when it comes to signs and symptoms that they exhibit. While considered as very rare cases, pediatric strokes are also considered as emergency situations, regardless if they are ischemic or hemorrhagic in origin.

Pediatric stroke can be divided into two categories, namely, perinatal pediatric stroke which occurs from the last 18 weeks of the gestational period up to the first 30 days of life, and childhood stroke which affects children with ages from 1 month up to 18 years of age.

The Cerebro Standstill

Risk factors for pediatric stroke include newborn infants, especially who are full term, older children who have other health conditions such as hemophilia, sickle cell anemia, congenital heart disease, and autoimmune disorders, and otherwise, healthy children with hidden medical conditions such as narrowing of the blood vessels or thrombosis.

Chapter 3 - Signs And Symptoms Of Stroke

The onset and progression of stroke depends upon its type. The onset of ischemic stroke may be sudden with gradual neurologic changes. On the other hand, hemorrhagic stroke often presents with a sudden bout of headache that's sharp, excruciating, crushing, and unbearable and may be accompanied with vomiting, often leading to a

The Cerebro Standstill

decreased level of consciousness or maybe even going into a coma. In both cases, the signs and symptoms of stroke are largely similar with a decrease in the level of consciousness, numbness or paralysis on one side of the body, difficulty in articulating words or understanding words, memory loss, and behavioral changes.

The signs and symptoms of stroke also depend on the area where the stroke occurs. When the stroke occurs in the frontal lobe of the brain, the person may have memory problems or have difficulty in understanding. When the stroke occurs in the temporal lobe of the brain, the person may have difficulty in picking up or listening. When the stroke occurs in the occipital lobe of the brain, the person may experience difficulty in seeing as this is the visual center of the brain. When the brain stem is affected by the stroke, the person may display a variety of signs and symptoms like a decrease in blood pressure, an abnormal heart rate, cessation of breathing, an increase in temperature, and more and oftentimes, strokes that affect the brain stem are considered as a top emergency case since this can lead to the death of the patient if not treated immediately.

The sudden onset of facial paralysis or weakness, arm drift, and abnormal speech are the key signs that a stroke may have occurred in a person. When all three of these are absent, the likelihood of a stroke is highly declined and other causes may have been the culprit in the neurologic deficits of the patient. While these characteristics may not be the perfect tool to diagnose stroke immediately, they are still useful in identifying stroke and they are valuable in rapid assessment and beginning treatment immediately.

In fact, some systems have been developed by the Department of Health of UK and the American National Stroke Association. These are the Los Angeles Prehospital Stroke Scale (LAPSS) and the Cincinnati Prehospital Stroke Scale (CPSS) which are useful in the rapid identification of patients who are suspected to have a stroke.

The signs and symptoms of stroke can be divided into various groups and subtypes, depending on the area that the stroke has affected the brain.

A stroke affecting the central nervous system (CNS) pathways which include the spinothalamictract, the corticospinaltract, and thedorsal column may include the following signs and symptoms:

- Hemiplegia or weakness on one side of the body
- Weakness on one side of the face
- Numbness
- Reduced motor or sensory function
- Muscle flaccidity
- Muscle spasticity
- Excessive reflexes

The Cerebro Standstill

Signs and symptoms of stroke often affect only one side of the body depending on which side of the brain the stroke has occurred and the opposite side of the body shows the signs and symptoms of the stroke while the same side of the face shows the signs and symptoms of the stroke. For example, if the stroke occurred on the left side of the brain, paralysis or muscle weakness may be felt on the *right* side of the body while there may be facial weakness on the *left* side of the face.

Aside from the CNS pathways, the brainstem may also be affected by the stroke and can produce a variety of signs and symptoms that primarily affect the functions of the twelve cranial nerves.

These signs and symptoms may include the following:

- Altered sense of smell
- Altered sense of taste
- Altered sense of hearing
- Altered sense of vision
- Decreased gag and swallow reflexes
- Limited pupil reactivity to light
- Drooping of the eyelid
- Weakness of the ocular muscles
- Problems in balance
- Nystagmus
- Facial weakness
- Muscle weakness
- Decreased sensations
- Paralysis
- Altered respiratory rate and breathing pattern

The Cerebro Standstill

- Altered heart rate
- Inability to turn the head to one side
- Inability to stick out or move the tongue

The cerebral cortex may also be affected by the stroke and the CNS pathways can also be affected, potentially producing the following signs and symptoms:

- Dysarthria – difficulty in articulating words
- Apraxia – altered voluntary muscle movements
- Expressive aphasia – difficulty in expressing words but retains ability to understand words such as reading and through verbal comprehension

(stroke affecting the Broca's area of the brain)
- Receptive aphasia – difficulty in understanding words like in speech or through reading, but remains the ability to express words through speech or in writing (stroke affecting the Wernicke's area of the brain)
- Global aphasia –combination of both expressive and receptive aphasias when a stroke has affected both the Broca's area and the Wernicke's area
- Defect in the field of vision
- One sided neglect
- Memory problems
- Confusion
- Hypersexual gestures
- Difficulty in concentration
- Lack of insight of one's disability

The Cerebro Standstill

When the stroke affects the cerebellum, a person may exhibit signs and symptoms associated with balance since the cerebellum controls balance and coordination of the body wherein in problems like the following may be seen:

- Abnormal walking gait
- Difficulty in coordination of movement
- Difficulty in fine motor movements
- Vertigo
- Difficulty in standing

Other signs and symptoms associated with stroke also vary. Depending on the onset, progression, and area of the stroke, the following signs and symptoms may or may not be present:

- Decreased or loss of consciousness
- Headache
- Nausea
- Vomiting
- Loss of bladder control
- Bowel incontinence

Chapter 4 - Diagnosing Stroke

With the variety of signs and symptoms that stroke can exhibit, it's not that difficult to establish a diagnosis. The primary goal when it comes to the diagnosis of stroke is to pinpoint its causes and the area of the brain where it is affected. A number of treatment plans and medications to be used for patients with stroke depend upon various diagnostic tests which may include the following:

- Complete blood count (CBC)
  o A CBC is used to determine if the blood clots too much or if there is the presence of infection through an increase in the white blood cells (WBC), as well as determining if there is adequate oxygenation through the number of red blood cells (RBC) in the blood which can decrease in the case of hemorrhagic stroke.

- Serum electrolyte studies
  o Routine serum electrolyte studies for sodium and potassium are often done to serve as baseline data.

- Serum sugar analysis

The Cerebro Standstill

o   Serum sugar can also be analyzed to determine the presence of diabetes mellitus in some stroke patients since diabetes is one of the most important risk factors when it comes to the development of stroke, both for ischemic and hemorrhagic stroke.

- Serum cholesterol analysis

o   Determining a patient's serum cholesterol levels, HDL, LDL, and lipid profile is important to know the presence of excessive fatty deposits that may have developed in the blood vessels and may be the cause of the stroke.

- Serum creatinine and BUN

o   A serum creatinine and a BUN analysis are often used to determine the function of the kidneys, but in cases of stroke, it is often used as a baseline to know whether or not certain drugs can be administered like diuretics (e.g. furosemide, mannitol, etc.)

- Electrocardiogram
o When a stroke affects a certain part of the brain (ex. brainstem), a patient may display an abnormal heart rhythm which can become alarming since the perfusion rate to the brain cells may not be the same.

- Computed Tomography scan (CT-scan)
o Using a series of X-ray scans on the brain, a CT-scan can detect a tumor growth, area of bleeding, stroke, and

The Cerebro Standstill

other conditions which make this diagnostic test an important examination when it comes to determining stroke.

- Magnetic Resonance Imaging (MRI)

o Another visualization exam like CT-scan, an MRI is like the more powerful version of a CT-scan as it can determine the areas where the stroke has affected and a dye may be used to highlight or visualize the blood flow to the brain and see areas of abnormality.

- Carotid ultrasound

o The carotid arteries may have an excessive buildup of fatty deposits or there may be the presence of a thrombus, making it a useful diagnostic exam for those with a suspected case of stroke.

- Cerebral angiogram
  - In this diagnostic examination, a small incision is made through the groin where a thin and flexible catheter is inserted which then enters the arteries and into your carotid or vertebral arteries; a dye is then injected into the blood vessels to visualize the blood flow to the brain and neck, detecting areas of abnormalities like ischemia, hemorrhage, tumor growths, aneurysms, and more.

- Routine physical examination

The Cerebro Standstill

o A routine physical examination is one of the strongest diagnostic tests which can determine the presence of stroke in patients even though it's the most basic test to be done upon entering the emergency room; various motor and sensory functions are tested, as well as reflexes which are big factors in determining the extent and progression of a stroke.

Chapter 5 - Management of Patients Stroke

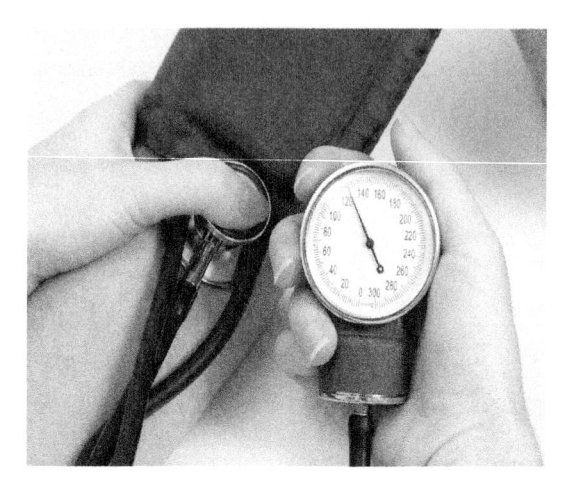

When it comes to the management of stroke, the treatment options depend upon the type of stroke and the area of the brain that is affected by the stroke. Oftentimes, supportive interventions are also implemented to prevent complications that are brought about by the disability that stroke can bring. During the recovery phase of a stroke patient, rehabilitation is done to recover what was once lost and promote self-care through physical rehabilitation, occupational therapy, and exercise.

Medical Management

Medications are primarily given to prevent increased intracranial pressure (ICP) in stroke patients, prevent vasospasm of the blood vessels, protect the brain tissue from further damage, prevent blood clot formation, and control blood pressure. With the variety of drugs given to stroke patients, strict monitoring is implemented to these kinds of patients.

Dissolve Blood Clots:

- Recombinant tissue plasminogen activator (rtPA) is used in dissolving blood clots through thrombolysis that may have been the cause of an ischemic stroke; however, this type of treatment is not used in hemorrhagic stroke as it will only worsen the bleeding.

Prevent Clot Formation:

- Antiplatelets prevent the platelets from sticking together and forming clots which further increases the risk of another stroke, but is not used in cases of hemorrhagic strokes.
   - Clopidogrel
   - Aspirin

- Anticoagulants prevent the formation of blood clots and prevent already existing blood clots from increasing in size, but are not used in cases of hemorrhagic strokes.

- Warfarin
- Heparin

Control Cholesterol Levels:

- CoA Reductase Inhibitors decrease the "bad cholesterol" or the LDL which limits the risk of the formation of fatty deposits in the blood vessels which can contribute to stroke.
  - Atorvastatin
  - Simvastatin
  - Lovastatin

Protect the Brain Tissue:

- Cerebroprotective Drugs protect the damage of brain tissue when there is an active phenomenon that can harm the brain like in stroke and head trauma.
  - Citicoline

Control Blood Pressure:

- Calcium Channel Blockers help in decreasing blood pressure, prevent chest pain, and prevent cerebral vasospasm.
  - Amlodipine
  - Felodipine
  - Nimodipine

- Diuretics not only help in decreasing the blood pressure, they also help in preventing fluid retention and osmotic diuretics are also helpful in preventing the increase in ICP in stroke patients.
  - Mannitol
  - Furosemide

- Beta Blockers help decrease blood pressure by blocking norepinephrine

The Cerebro Standstill

and epinephrine to bind from their receptor sites.
- o Metoprolol
- o Bisoprolol

Other Medications:

- Laxatives can be given to prevent straining which can contribute to an increase in blood pressure (e.g. lactulose, bisacodyl maleate).
- Multivitamins and appetite stimulants may also be given as supplementary support for the nutrition of stroke patients.
- Anti-ulcers are given to prevent the formation of gastric and duodenal ulcers, especially if a stroke patient has not been allowed to eat per orem for an extended period of time.

- Analgesics may also be given in cases of severe pain like headaches or joint pain for stroke patients.
- Antipyretics are given in cases of fever that may be due to the changes from the brainstem or an infection brought about by prolonged disability.
- Antibiotics may also be given to prevent infection or to treat an active infection that may have developed throughout the course of the stroke, like cases of pneumonia which are highly common for stroke patients.

Surgical Management

Surgery may also be done for stroke patients through craniotomy or through hemicrainectomy, which is the removal of one side of the skull to evacuate the accumulated blood and relieve the excess intracranial pressure, as in cases of hemorrhagic stroke.

Angioplasty and stenting are also surgical options that may be done to treat an ischemic stroke and prevent future stroke attacks. Aneurysm clipping may also be done to prevent the worsening of a detected aneurysm and prevent its rupture.

Surgical management of stroke primarily depends on the extent of the stroke and the areas of the brain that are affected by the stroke. When medical management seems to be ineffective, surgical treatment is looked as an option, but otherwise, medical intervention is the primary mainstay treatment option for both ischemic and hemorrhagic stroke.

CONCLUSION

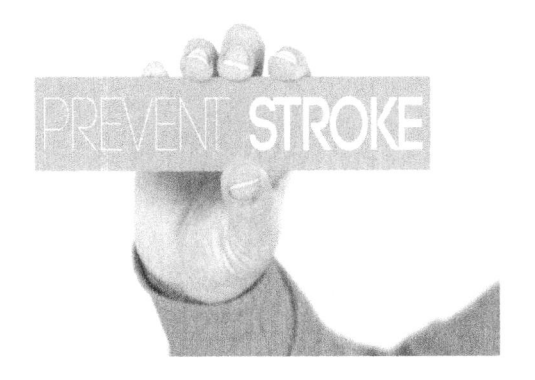

The brain is a powerful organ that governs every part of our body, even the part that we can't even control. It's positioned on top of our heads, signifying its rule over the body. However, just like with other organs, it's far from being omnipotent. Once it doesn't get adequate nutrients and a decent supply of oxygen, the brain can get damaged and the brain cells can even die, resulting in irreversible neurologic changes in people.

With the rise in the incidence of stroke, understanding its causes, risk factors, signs and symptoms, and management is important. While it's certainly alarming that the number of people experiencing stroke is on the rise, knowing about this condition is essential to protect yourself, your family, and your loved ones.

Both ischemic and hemorrhagic stroke are certainly fearsome with the disability that they bring even after successful treatment. However, the most powerful weapon that we can get from this is that prevention is always the key factor in controlling these diseases. Proper exercise, proper diet, and proper lifestyle habits are key factors in preventing the occurrence of stroke in the first place.

The Cerebro Standstill

And hopefully, this Book has helped you shed light on one of the most debilitating and formidable medical conditions in the world.

Printed in Great Britain
by Amazon